Our Exciting Earth!

REEFS

By Tanner Billings

Gareth Stevens
PUBLISHING

Please visit our website, www.garethstevens.com.
For a free color catalog of all our high-quality books, call toll free 1-800-542-2595 or fax 1-877-542-2596.

Library of Congress Cataloging-in-Publication Data
Names: Billings, Tanner, author.
Title: Reefs / Tanner Billings.
Description: New York : Gareth Stevens Publishing, [2023] | Series: Our exciting earth! | Includes index.
Identifiers: LCCN 2021039095 | ISBN 9781538275726 (Set) | ISBN 9781538275733 (LibraryBinding) | ISBN 9781538275719 (Paperback) | ISBN 9781538275740 (eBook)
Subjects: LCSH: Reefs–Juvenile literature. | Coral reefs and islands–Juvenile literature. | Reef plants–Juvenile literature. | Reef organisms–Juvenile literature.
Classification: LCC GB461 .B55 2023 | DDC 577.7/89–dc23/eng/20211008
LC record available at https://lccn.loc.gov/2021039095

First Edition

Published in 2023 by
Gareth Stevens Publishing
29 East 21st Street
New York, NY 10010

Editor: Kate Mikoley
Designer: Tanya Dellaccio

Photo credits: Cover V_E/Shutterstock.com; p. 5 silvae/Shutterstock.com; p. 7 Damsea/Shutterstock.com; p. 9 Dobermaraner/Shutterstock.com; p. 11 SARAWUT KUNDEJ/Shutterstock.com; p. 13 Pawe? Borówka/ Shutterstock.com; pp. 15, 19 Volodymyr Goinyk/Shutterstock.com; p. 17 stephan kerkhofs/Shutterstock.com; p. 21 tororo reaction/Shutterstock.com; p. 23 ChameleonsEye/Shutterstock.com.

Printed in the United States of America

CPSIA compliance information: Batch #CSGS23: For further information contact Gareth Stevens, New York, New York at 1-800-542-2595.

Contents

A reef is in the sea.
It is near the top.

Some are made of rock.

Some are made of
tiny animals.
These are coral.

Coral has a hard part.
It can form a reef.

Lots of fish live here.
Plants do too.

Corals are many colors.

Some are red.
Some are blue.

Reefs grow for many years.

They can be big.
The Great Barrier
Reef is big.

You can go to a reef.
Don't touch!

Words to Know

coral

fish

plant

Index